I0469436

How To Find, Choose and Buy Your First Telescope.

A Guide for Students and Parents

By: Keith Nichols
(c) 2012 Fivepenniespress

I have been practicing amateur astronomy for ten years now, mostly from a very light polluted area in western Washington state. Over the course of this decade I have learned, mostly through trial and error, the best and worst ways to buy a telescope, use a telescope, and set up a telescope. I have no formal astronomy training, except that which I can glean from books, the web, and our old friend Carl Sagan.

My own personal telescope collection at this time consists of:
An 8 inch Meade Schmidt-Cassagrain
An 8 inch Orion Dobsonian
A 4 inch Celestron refractor
A 4 inch Galileo reflector
A 70 mm Meade Go To refractor
A 4.5 inch Bushnell reflector
A 70 mm Celestron Powerseeker refractor

Plus several cheap Walmart special style 60mm refractors, which, although they are completely useless and generally the worst telescope being made now, I have a certain grim fascination with. I buy these steaming piles and collect them for no real good reason.

A very cold night of stargazing.

Buying a telescope is the first step in embarking on a rewarding journey into the world of amateur astronomy. Unfortunately its also where most people make the biggest mistake, which leads to massive amounts of frustration and eventually ends up with the buyer placing the telescope in a closet. If you buy a decent telescope to start out with, you will have a lot more fun and get a lot more enjoyment from your purchase. A cheaply made telescope will only create frustration.

I get most of my telescopes from craigslist, from people who buy them and never learn how to use them properly. They then put the scope in the closet, in the basement, or in the garage and forget about it. This happens to good quality scopes as well as cheap telescopes. Because of this, craigslist is a good place to find a nice telescope for a very reasonable price.

First though, lets discuss the kinds of telescopes, and figure out which one you want. Researching the market and buying the scope itself will come after you have figured out what will work best for you.

This book is for the beginner, not for the experienced amateur, so if you own a dozen telescopes, spend long cold winter nights taking amazing photographs and discovering new comets and such, this information might not be new to you at all. You may consider it boring in fact. But hey, we are all in this together, so lets get started.

Make some coffee, and take notes, there will be a quiz at some point.

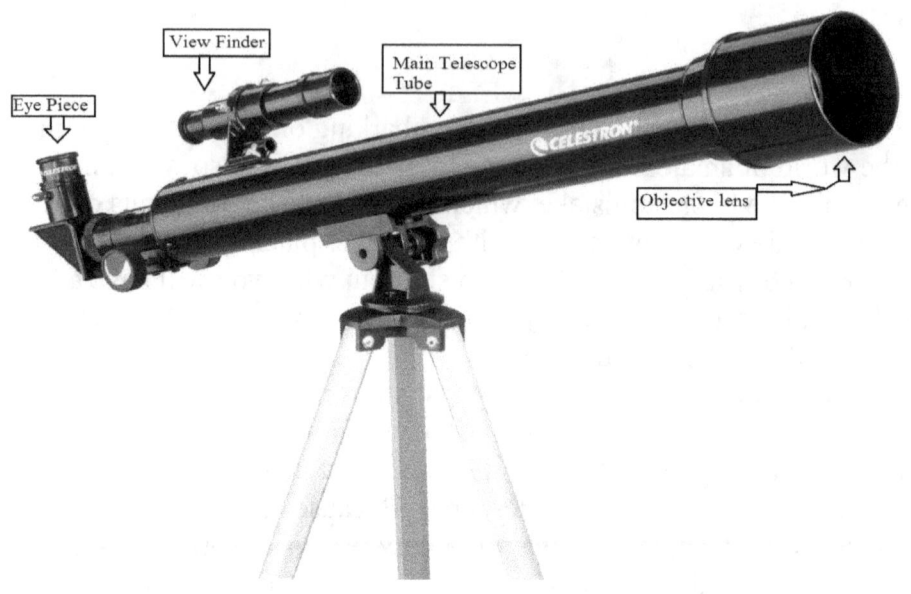

Galileo's first telescope was a small refractor, meaning it used a series of lenses to collect the image. Everybody is very familiar with the refractor telescope. This image is of a simple refractor telescope.

It is also a very poor quality telescope, and I certainly do not recommend this scope for beginners at all. But it works good for illustrative purposes. You can see at the right hand side the main objective lens, then following down towards the left, you have the telescope tube, focusing rack and pinion, and finally the eyepiece. The small tube attached to the top of the main telescope is the view finder scope, which you use to locate the object you wish to view through the telescope. This is the basic layout all refractor style telescopes will follow. No big surprises here, very simple, but also a very easy way to build a scope.

Next is a reflector style telescope, sometimes called a Newtonian, since Isaac Newton invented it. This following illustration is a reflector.

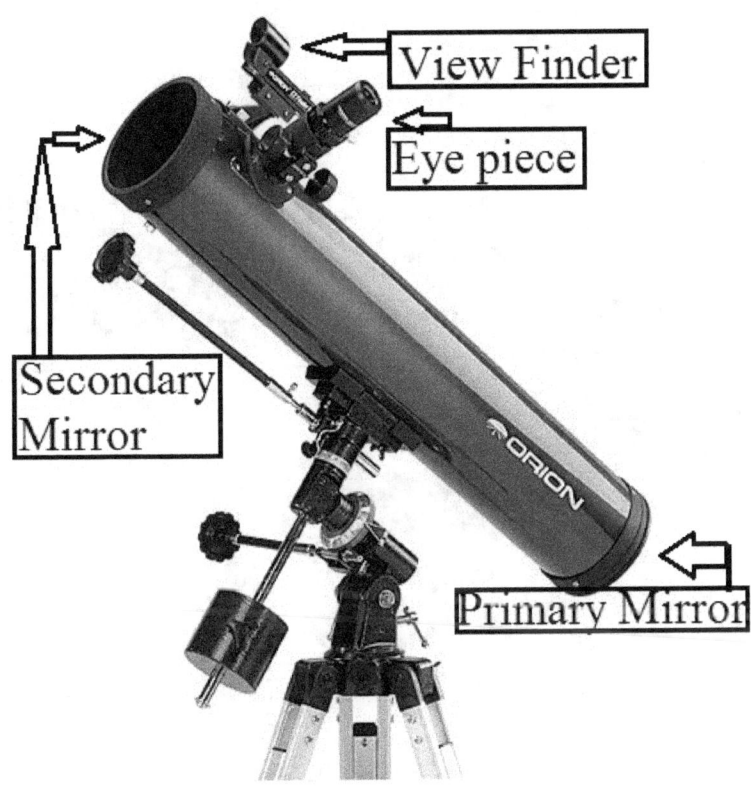

This telescope is a pretty good scope, one that I would recommend for a beginner to use. As you can see from the illustration there are many differences between this and a refractor style telescope, and the two styles are pretty easy to tell apart. A reflector uses a series of mirrors to collect the image, instead of lenses. A big main mirror, or primary mirror is located at the base of the telescope tube, and a smaller secondary mirror is located at the top of the tube. Light is gathered by the primary, reflected up to the secondary, and reflected from there into the eyepiece.

There are several variations on these two basic themes, but this book is for the beginner, not the seasoned veteran. Either one of these two styles of telescope, a refractor or a reflector, would work great for the budding astronomer. And both are very easy to use, and simple to set up. It all depends on the actual scope you purchase.

Lets look at the image I used for the refractor, the cheap telescope that most people are familiar with.

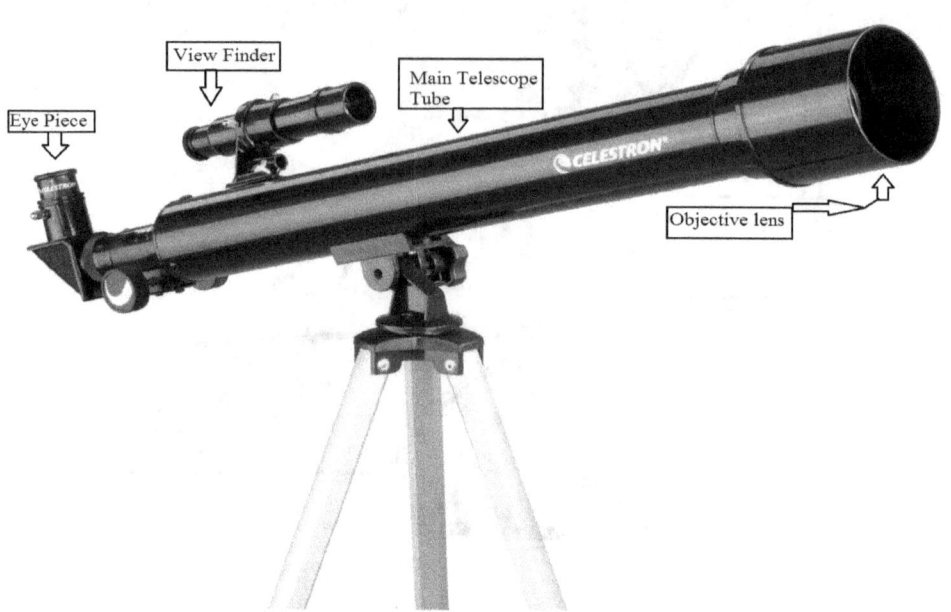

This telescope is very poorly made, and will cause nothing but headaches. They are cheap, and that's about the only thing it has going for it. I believe Walmart sells these around the holidays for about 35 dollars. Which seems like a great deal, and they look pretty cool under the Christmas tree. However, when your 14 year old daughter sets it up later in the evening, and tries to actually use it as a telescope, she will soon learn why it was 35 dollars. First off the tripod is very unstable, and will move, shake, rattle, and vibrate at the slightest touch. The mount, called an Alt-Azimuth style mount, is shaky, hard to control, and will need constant adjustment. Which, of course will cause more vibration on the image seen through the scope. The eye pieces are often cheaply made plastic, usually of a one inch diameter. These are not usable in the better quality scopes, and work only with these cheapies. (All high quality telescopes use 1 ¼ inch eyepieces at least.)

The main objective lens in these telescopes are poorly made, with no coatings usually, and will scratch and cause a lot of image aberration. (meaning the image itself, when you finally get the scope to sit still long enough to look through it, will be distorted and have a false color to the edges.)

At the very best, in very good conditions, with hours of patience, she might be able to view the moon with this. But your daughter will soon grow tired of the wobbly nature of the scope, the poor mount and tripod, and will loose interest fairly quickly.

Do not buy these telescopes.

The absolute worst telescope I have ever seen is this one.

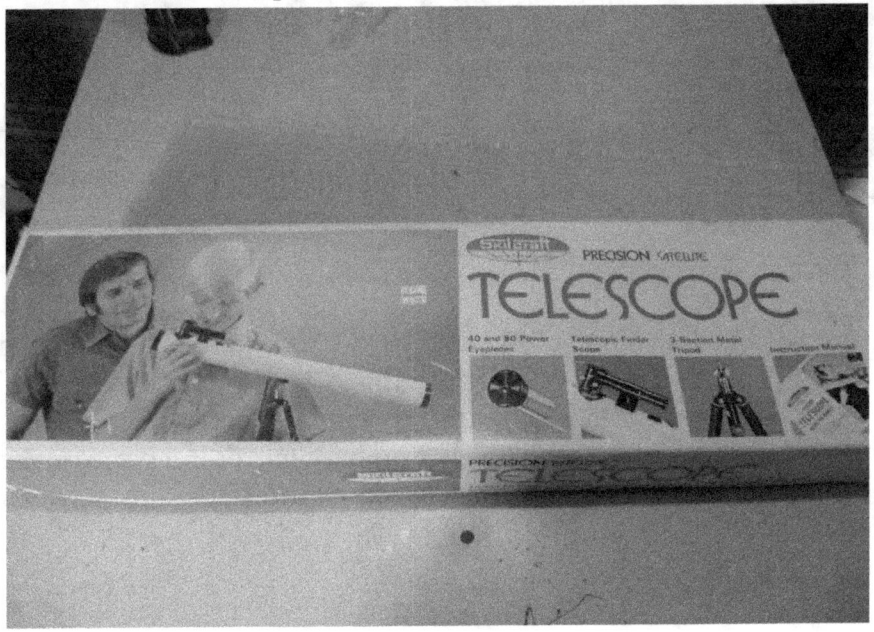

I bought it at Goodwill, simply because it was so bad I couldn't resist. I am a sucker for things like this, just ask my long suffering wife.

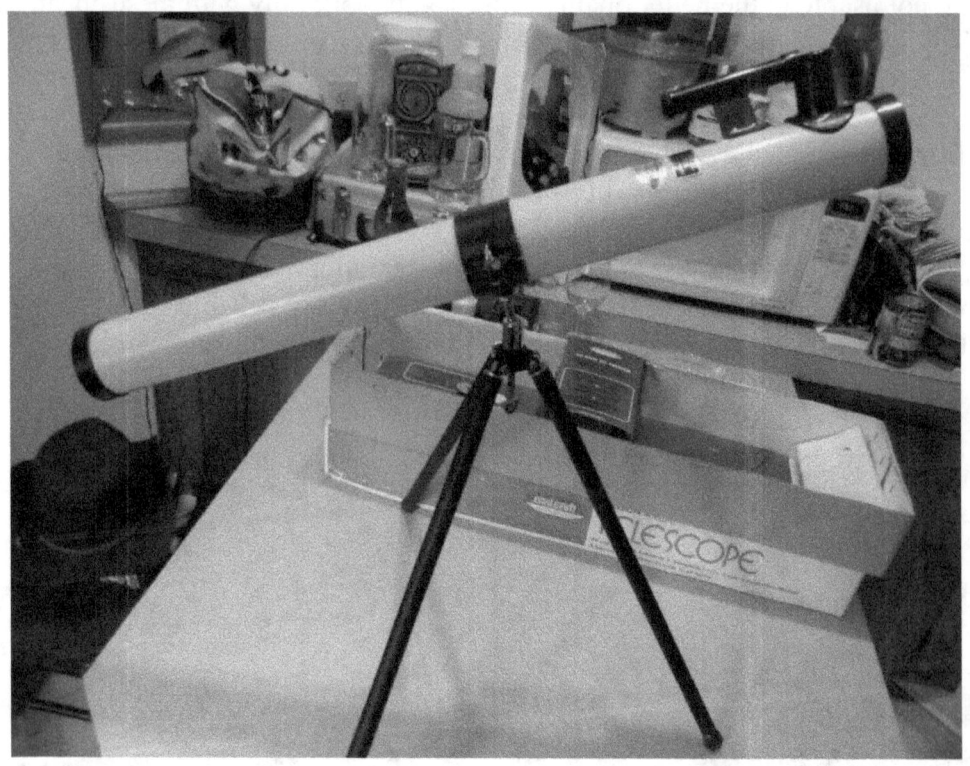

Here it is in all it's finite glory. Its a 50 mm reflector, made from a piece of white PVC pipe, mounted on to a very flimsy tripod with a wing nut. The eyepieces that came with it are little bitty things made from cheap plastic, with cheap plastic lenses in them. The primary mirror is a piece of aluminum with a shiny coating on it, and the secondary mirror it a shiny sticker stuck to a bit of plastic.

It doesn't even have a focusing knob, to focus this telescope you pull the eye piece in or out of the tube.

The complete worst thing I have ever paid 10 dollars for.

Let's figure out what kind of telescope is best for your needs, then I will show you some better alternatives.

Refractor or reflector?

Both styles have their advantages and their disadvantages, the main advantage to a refractor telescope is their relative light weight and ease of use. The main draw back with these is the price, the bigger the aperture, (size of the lens) the more expensive they are.

A reflector telescope is not as expensive, but is heavier, and requires a slightly higher learning curve to set up and use. However, the bigger aperture of a reflector relative to its initial investment is much lower.

Both style of telescope require no maintenance, in fact cleaning a telescope lens or mirror is the best way to destroy a telescope. (I know, I have tried.)

What are you going to use the telescope for? Are you going to be taking it out once a month, to gaze at the craters of the moon perhaps? Or are you planning on learning how to take photographs through the telescope and doing some serious planet viewing? Or do you think your son would enjoy just sitting in the yard and viewing stars and star fields? All of these things need to be considered before you make your purchase.

While all styles of telescope will enable you to do all of these things, some of them are better suited to specific tasks, like say viewing the planets, while others are best at looking at the moon.

For the beginner, the complete noob at astronomy, I recommend either a good quality 4 inch reflector, or a good quality 90 mm refractor.

A 4 inch reflector will do decent planet viewing, be inexpensive, and fairly easy to use. While a 90 mm refractor will give great views of star fields, and distant objects.

Both will give good viewing of the moon.

Mounts

There is one more thing to consider before we look at some scopes. And that is the mounting style of the telescope. How it attaches to the tripod is pretty important, as the various ways of doing this will enable you too either get a good solid viewing platform, or a wobbly hard to use platform.

The tripod itself is also pretty important, although most good quality scopes come with a good tripod.

There are a few basic styles of mount to consider.

The Alt-Azimuth Mount.

The Dobsonian Mount

The Equatorial Mount.

The Go To Mount.

And of course all of these mounts have combinations. You can get a go to equatorial mount for example.

The simplest design is an Alt-Azimuth mount, which in its most basic form is a U-shaped bracket on a swivel, which will allow you to move the telescope into any position you desire, and lock it down once you get there with a thumb screw of some kind. While these mounts are very inexpensive and are often on a beginner style telescope, I do not recommend them for the novice, as they are rather difficult to get stable and will not give a pleasant nights of viewing.

This is a simple Alt-Azimuth style mount.

It is not a very stable mount, and if you will notice, the telescope in the image is a fairly cheap and poorly made one.

What you want in a mount is stability, because nothing causes frustration faster then an image of the moon that won't hold still. Or finally getting Saturn in your field of view only to accidentally bump the telescope and have it go bouncing off in some random direction and loosing your view completely. Saturn is a very small target relative to the rest of the night sky, and loosing it in you field of view can be very frustrating.

A better style of mount, one of my favorites, is an equatorial mount. An equatorial mount will enable the telescope to track an object in the sky, either manually with a scroll wheel type control, or electronically with a drive motor. They are usually pretty well made and give a nice solid platform for viewing. They are slightly difficult to set up, but once they are set up properly, the telescope will be much more enjoyable. Lets look at an equatorial mount.

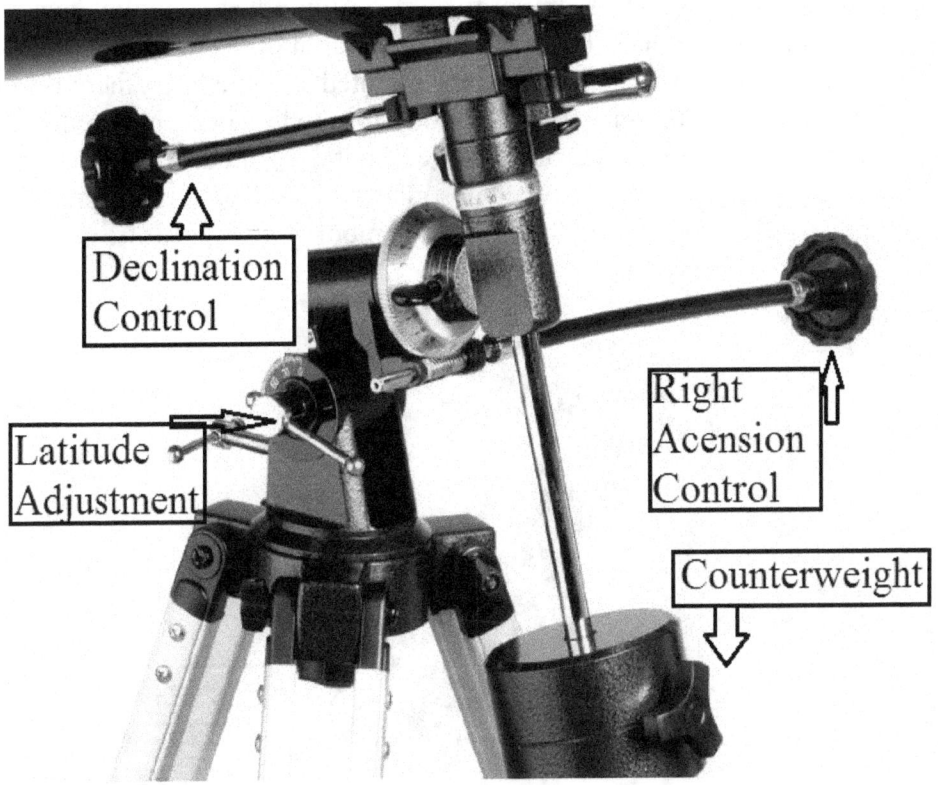

Declination Control

Latitude Adjustment

Right Acension Control

Counterweight

This is a decent, inexpensive, and manually operated equatorial telescope mount. The Counterweight is a large weight that offsets the weight of the telescope, resulting in a perfectly balanced unit when all set up. Right acsension control moves the telescope in a panning style of motion, while the declination control adjusts the telescopes angle in relation to the horizon. The latitude adjustment is set for your areas latitude, say 45 degrees in the northwestern united states. The mount is set up on a level tripod, with the Right Acension barrel of the mount pointing towards due north. Once you find the object, by moving the telescope on the mount, and you lock the gimbals or thumbscrews into place all you need to do is adjust the right acsension and the telescope will follow the object.

A motor drive, attached to the right acsension of such a mount will allow hands free tracking of an object.

A Dobsonian style mount is similar to an Alt-Azimuth style mount, except for a few design differences. First off, a Dobsonian mount is very stable. Once a telescope on a Dobsonian mount is pointed at something, there is very little vibration. These mounts are pretty good for the beginner, but are heavy, and by their nature are very simple. They will not track an object, unless you pay for a very expensive go to style Dobsonian mount. But they are pretty good for star field viewing and looking at the moon.

This is a 4.5 inch reflector on a Dobsonian mount. As you can see, its a very stable and very simple design. They lend themselves well to larger reflector telescopes. The load spring acts as the counterweight on this mount replacing the heavy weight on the equatorial mount. The best thing about a Dobsonian mount is its ease of use, literally just set the scope in the yard and you are ready to go. This is a very good telescope, one I would recommend.

A Go To telescope can be mounted on any one of these mounts, and because of its computer control, it will track an object, and go to it for you with its drive motors. Go To scopes are very nice for the beginner, but there cost is much higher then a manually operated telescope. The biggest drawback to a Go To style mount is the lack of learning, in my opinion. They will find things very easily, but they wont teach you or your daughter how to find them. Their expense makes them pretty far out of reach for the beginner.

If you purchase them used, which we will get to, then you can get them for a reasonable price and they make an attractive option.

This is a very good telescope. The drawback with this particular one is the Go To alt-Azimuth mount will cause the object viewed to rotate, making it not a good choice for taking photographs.

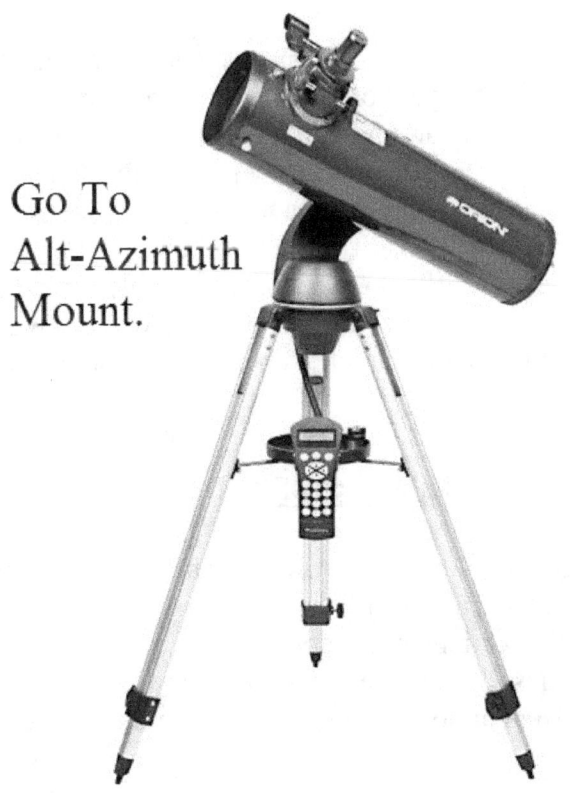

Go To
Alt-Azimuth
Mount.

So we have covered the basic styles of telescopes, and the basic kinds of mounts for them. We are making pretty good progress and you are doing a good job of keeping up. How are the notes coming along? Looking good? Good job. Have a cookie break and we will get to the brass tacks.

Buying a telescope.

Now that you know what you want, the styles of telescope to choose from, and the kinds of mounts available, we can start looking at telescopes to purchase.

There are two ways to buy a telescope, new or used.

The best place to get a new telescope is through amazon.com or a local store that sells them. But amazon is best I think because of the wide variety of makes, models, and styles to choose from. And with the shipping options available, you can have your scope within a day or two.

When buying used, the best place is craigslist.com. Just search telescope, title only, with your price range. For a beginner scope, expect to pay around 100 dollars on craigslist, and around 150 on amazon, ignore the super cheap telescopes because, as we have learned, they just aren't worth the hassle, headache, and the money. When buying a telescope, you get what you pay for. A 45.99 special on amazon isn't going to be a very good scope, while a 299 scope, marked down for the holidays or just on sale for 149 is a much better option. Shop around, take your time, and you will see what I mean.

Sometimes you can find a decent telescope for free just by asking around your family.

New Telescopes

New is best, by far. You don't have to worry about a dirty mirror or lens, or a broken mount. You have a warranty backing up the product, and you get the instructions for the particular telescope.

My recommendations for beginner telescopes, new on amazon.com, are as follows. These are listed in no particular order, with no preference to any one brand or manufacturer.

The Orion Space Probe 3 Equatorial telescope.

This scope is a high quality, easy to set up and user friendly manually operated equatorial reflector telescope. It has a 3 inch primary mirror, not super big, but also not too heavy.

At the time of this writing it retails for 159.99 and was on sale for 129.99. Good buy for a good quality beginners telescope.

The Celestron Powerseeker EQ 70 MM

This is a good 70 mm refractor telescope, on a decent manually operated equatorial mount. The optics on the scope are high quality, the focuser works very smoothly, and once its set up properly, the EQ mount is pretty stable.

It retails as of this writing at 171.99 and was on sale for 98.99. Pretty good price, and a good telescope for under a hundred dollars. I personally own one of these and a 4 inch Celestron refractor. High quality telescope.

And that's it, really, as far as affordable, new, high quality telescopes for beginners. There are others to choose from, for sure, but the good ones are all just variations of these two. 70 mm refractor, or a 3 inch reflector. Not a whole lot of choice, but they are both good scopes. However, if you buy a used telescope, then your options open up a lot.

Used Telescopes.

The biggest drawback to buying a used telescope is the lack of warranty and information when you make your purchase. Also you run the risk of getting a damaged or dinged up telescope. But if you shop around and check out the telescope closely before you commit any money to the purchase, I feel you can get a good quality telescope for a fraction of the cost of a new one. For examples, I have performed a quick search on craigslist, with the parameters I stated above. (telescope, title only, 5-150 dollars)

I found several quality instruments for sale, all of which are suitable for a beginner, and all of which fall between the price range of 5 to 150 dollars. Of course, I didn't find any good ones for 5 dollars. Don't be silly.

Here are some examples:

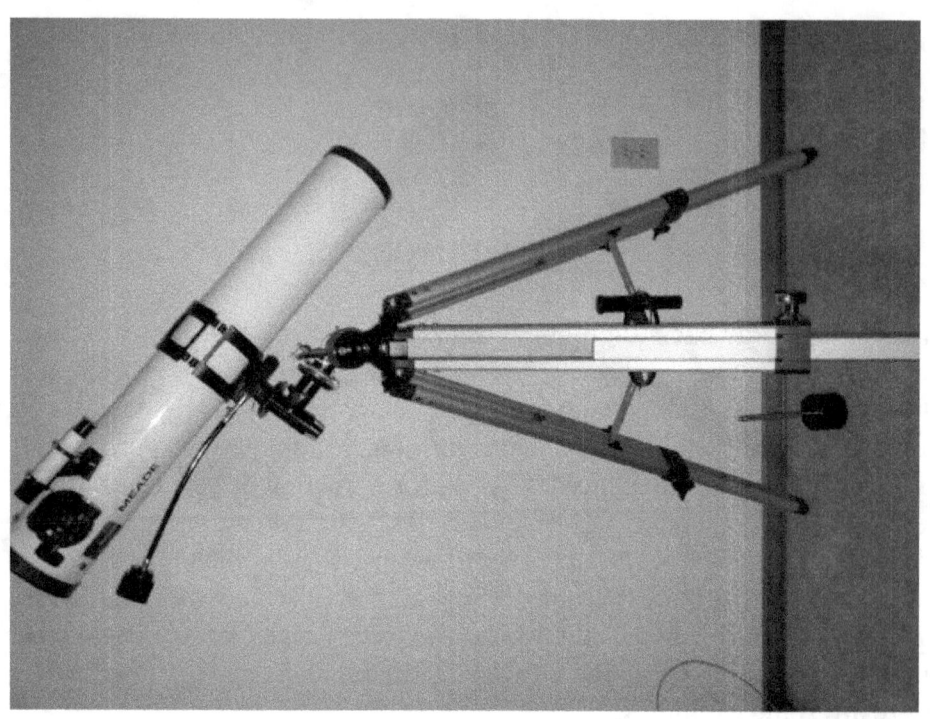

This is a Meade 4.5 inch reflector on what appears to be a good tripod with an equatorial mount. All the controls are there, it includes 3 eye pieces and all the dust covers are there. The counterweight is down on the floor by the tripod. They were asking 40 dollars cash for this. Pretty good buy I think, definitely worth a look at.

This telescope is a Stratus brand, kind of an off brand I suppose. It is a 4 inch reflector, also on an equatorial mount. They were asking 75 dollars for this one, however the Meade pictured above is 35 dollars less and is the better telescope of the two.

This is a Meade Go To 70 mm refractor. It included a couple eyepieces, and looks to be complete. They were asking 100 dollars for this telescope, which seems like a good buy. I would take a look at it for this price, it is a go to scope and we have discussed the high price of these new.

Of course a real good way to tell how much a scope is actually worth on craigslist is to find a comparable model on Amazon.

This is a Celestron 90mm Astromaster EQ. Good telescope, sitting on craigslist, for 100 dollars.

So you can find some pretty nice instruments just by doing a quick craigslist search. Its not hard to call somebody and set up a meeting in a public place and take a look at a telescope. I have bought several used scopes this way and all of them have been in like new condition. And they usually include extra eyepieces and other accessories. It isn't too difficult to end up with a dozen telescopes once you get started in the hobby. Here is an image of a few of mine.

All of these were purchased off of craigslist over a period of a year. My wife says I need to have an intervention.

Accessories

The most common accessory for your telescope is eyepieces. Think of the telescope as the hardware and the eyepiece as the software. The telescope is set, once you get it it made and purchased and set up, its power isn't going to change. The eyepiece is what gives the telescope its ability to change its magnification, or power. Some real cheap telescopes make outrageous claim at high magnification. Beware of these claims. While it is possible to get very high magnification from these, its pointless to do so since you wont be able to view the magnified image very well at all. The beginner in astronomy sometimes think they need a dozen or so different size eyepieces. This isn't true. All you require, really are three quality eye pieces and a Barlow lens. A Barlow lens increases the power of you your eyepiece, with by doubling it or tripling it. So a 2x's Barlow will double the power of your eyepiece, while a 3x's Barlow will triple it.

Eyepieces come in many sizes, and the smaller the size, the more powerful they are. So for example, a 16mm eye piece isn't as powerful as a 9 mm. The smallest you really need for a beginner is a good 9mm. I recommend at least these 3 sizes, a 9mm, 16mm and a 24mm. You can order on amazon an accessory kit that includes many sizes of eye pieces and a Barlow lens, as well as filters for them all. Its not really a needed accessory to have all these eye pieces though. As I said, the three sizes above, plus a 2x's Barlow will work fine. With the Barlow lens you effectively have 6 different sizes to choose from.

Other styles of telescopes

This book is intended for the beginner, not the serious astronomer. For somebody just starting out, a simple reflector or refractor is all you want and all you really need.

However there are a few other styles out there that I feel I should mention, to be complete.

A Schmidt-Cassagrain telescope is a very high end instrument, and very expensive. I own a Meade 8 inch SC, and its my main telescope I use. They combine a reflector with a refractor. A Schmidt-Cassagrain uses a large

primary mirror, a smaller secondary mirror, and a large corrective lens over the front of the telescope. This design results in a very compact layout, but with very high quality image capabilities. Here is a nice SC telescope. It is a Celestron Nexstar 8 se. Fully go to and computer controlled, this scope will do anything you want it to. Expect to pay about 2000 dollars for a scope like this, and of course the bigger the scope, the higher the price. My personal dream telescope is this Meade.

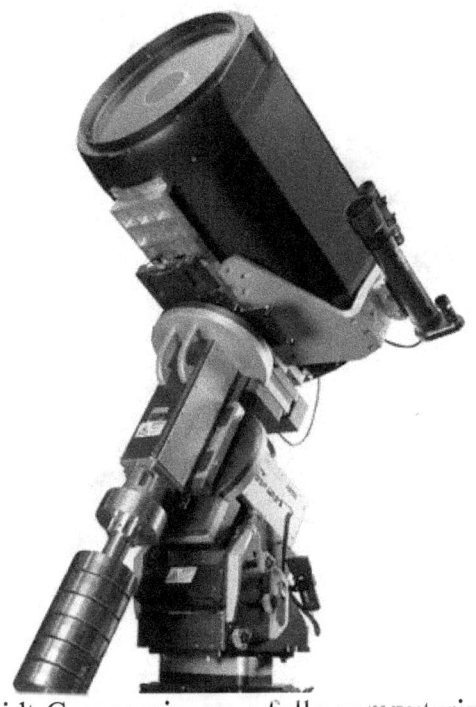

Its a 15 inch Schmidt-Cassagrain, on a fully computerized mount. This particular telescope sells for 35 thousand dollars. Not very well suited for the beginner. But very drool worthy. I believe this telescope will even bake you dinner.

Another style of telescope is a light bridge Dobsonian mount. Its basically the same as a reflector, but large and sometimes set up on a go to style Dobsonian mount. They are very nice large aperture telescopes, and aren't really all that horribly expensive. You can get a good sized light bridge for under a thousand dollars on amazon, and they show up on craigslist occasionally.

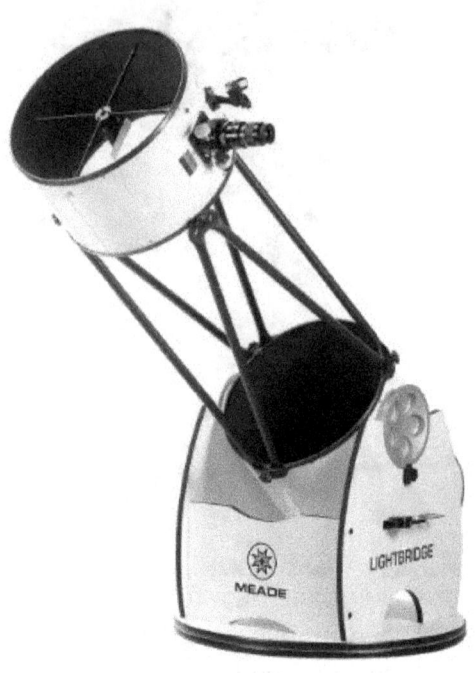

This is a nice Meade light bridge. As you can see, they are very similar to a basic Dobsonian mounted reflector, but instead of having a solid and heavy main tube, they have a trusses holding the secondary mirror and eyepiece assembly away from the primary mirror. This makes for a light weight design for a very large telescope. And they aren't too expensive, this particular scope is under a thousand dollars at 995.99.

So what can you see with your telescope?

You can see the moon, the major planets Mars, Mercury, Venus, Saturn, and Jupiter. You can see galaxies and star clusters, and many stars that will split into double or triple stars. I have been taking astronomy photographs pretty much since I started and these will give you some idea of what things will look like in the telescope. Keep in mind there is some image degradation from telescope to camera, the image you see in the scope will always be superior to the image captured on a camera.

All of these images are unedited and are not stacked. They are single frames of exposure.

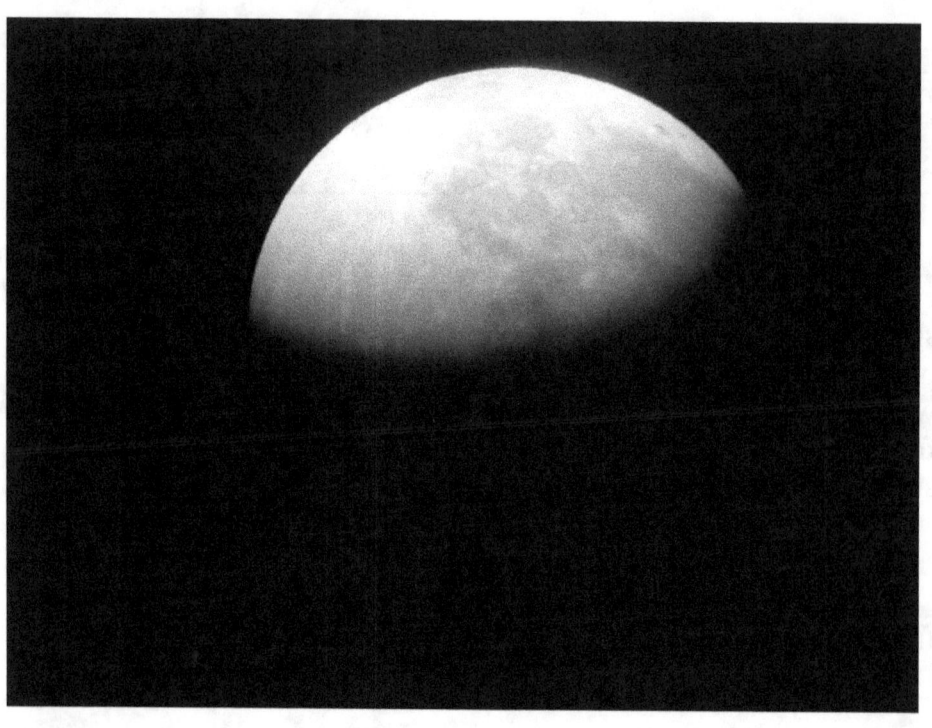

The Moon as seen through a 4 inch reflector with a low power eye piece.

The Moon through a four inch reflector with a slightly higher powered eyepiece.

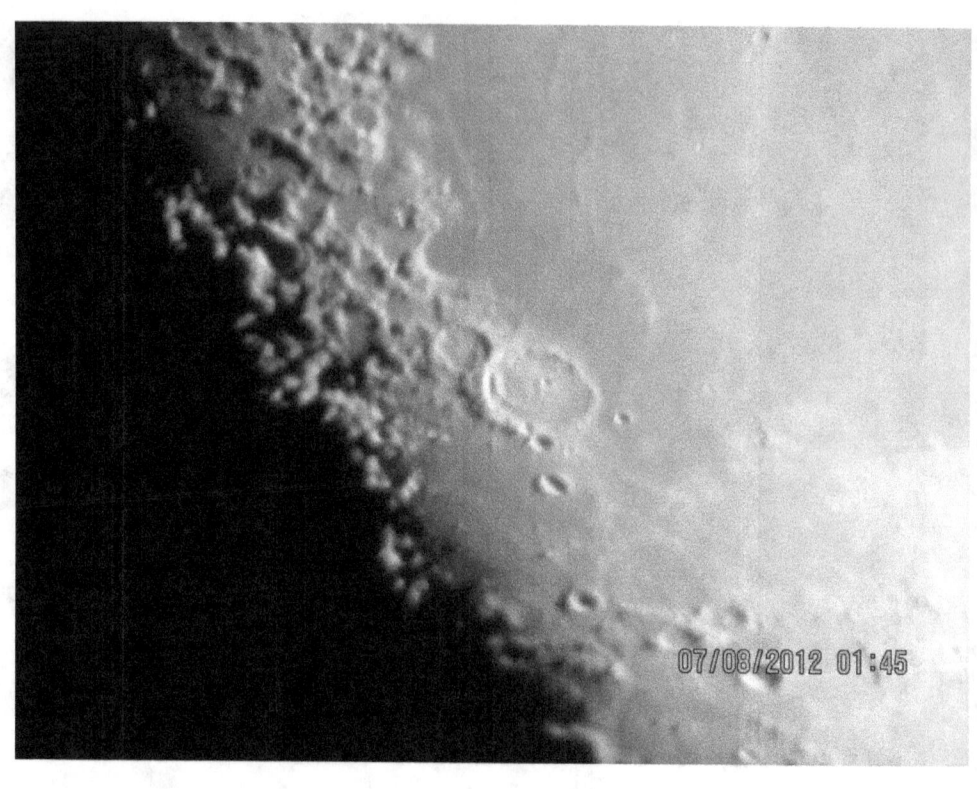

07/08/2012 01:45

The Moon at the limits of power with a 4 inch reflector. This was with a 9mm eyepiece and a 2x's Barlow lens.

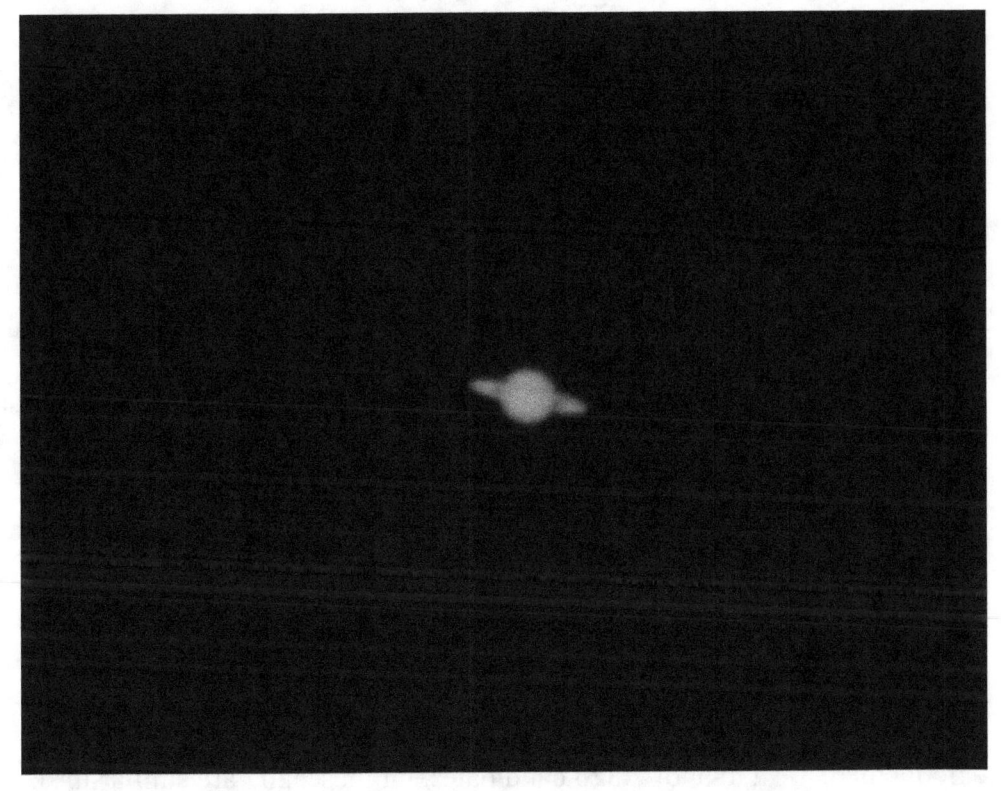

Saturn with a 4 inch refractor with a high power eyepiece.

Some closing thoughts on astronomy.

What was it that set me down this path? Of exploring things and wanting to always know more? There was one man who influenced me above any others. I consider him to be my teacher and my friend, although I never met him personally. I would have been extremely grateful to have met Carl Sagan, and to have shaken his hand and told him what he meant to me. Sadly I never had the chance.

If you have not seen Sagan's Cosmos series, go watch it, its available freely on YouTube and the DVD set is very comprehensive. It will teach you more then I ever can in the brief time I have here. I could write a dozen books on the subject, and none of them would even begin to compare to Sagan's collected works.

As I look at all this amazing stuff out there, I of course become more and more curious as to what it is and where it is and how it got there. Its a natural thing for a human to want to know more. The nature of the universe is vast, huge, and beyond me. Not that I don't understand it, I do, I grasp, barely, that is numbingly large and changing almost constantly I mean its beyond me because it literally is. Everything about the universe I want to know more about is beyond me by vast distances. Our nearest neighboring solar system, that around Alpha Centauri, is a mere 4 light years away. But that distance is so vast, at our present technological level, we would get there in some many thousands of years. But oh how I would love to go.

One of our telescopes we have in space right now is the Kepler observatory. Kepler's mission, its sole reason for being up there, is to look for extra solar planets, meaning planets that are outside of our solar system.

So far there are a total of 2,326 candidates. Of these, 207 are similar in size to Earth, 680 are super-Earth-size, 1,181 are Neptune-size, 203 are Jupiter-size and 55 are larger than Jupiter. Moreover, 48 planet candidates were found in the habitable zones of surveyed stars. The *Kepler* team estimated that 5.4% of all stars host Earth-size planet candidates, and that 17% of all stars have multiple planets. In December 2011, two of the Earth-sized candidates, Kepler 20c and 20e were confirmed as planets orbiting a Sun-like star, Kepler-20.

What is most interesting about these number is Kepler's field of view.

Kepler doesn't scan the entire sky, but instead focuses on a very small portion of the sky, peering intently at it for years at a time. How small?

About the size of your fist held up at arms length. Hold your fist up, and look at it against the back ground of the sky, or the ceiling I suppose. In the space of your fist, Kepler has found over 2,300 planets.

How can we possibly be so ignorant to think of ours as the only intelligence in the universe with a revelation like that staring back at us from the night sky?

We use steel to make may different things, from cars to forks, from ships to baking pans. And we very rarely even give it a second thought. I think about steel a lot, not just because of my steel art work, or my job of machining steel, but because its made, mostly from iron. You take some iron ore, and smelt it down with some carbon and various other additives and you get steel.

Iron is the 4th most abundant element in the earths crust. The core of the planet is made from molten iron, intensely hot, and responsible for our magnetic field, the changing geology of the planet, and our atmosphere.

I have asked some people where iron comes from and most say it comes from mines. Which to them is a perfectly reasonable answer. It does come from mines. We have to dig it up out of the earth to get to it. But that's not where iron comes from.

The iron in our steel comes from exploding stars. When you are holding a bolt or a piece of cast iron, or that fork, you are holding in your hand a piece of a dead star that exploded eons ago in a spectacular super nova explosion.

I find it amazing that we humans can hold in our hands a bit of a dead star and be ignorant and stuck in our arrogance and not understand where it comes from or what it is.

It is astounding to me that I can take my telescopes outside on a clear night and look at the rings of Saturn and the moons of Jupiter from my front yard. We tend to take these things for granted, assuming they are just there and will always be, yet we rarely if ever manage to look at them with our own eyes. With the telescopes I have recommended for you and some basic practice you will be seeing these things on your own. I hope when you or your daughter or son does, they will be as excited about this as I have become.

Astronomy isn't hard to do, but getting people interested in astronomy is very hard to do. If I wrote a book on Astrology, complete with signs of the zodiac and vague predictions about how mars would influence your life, I may have better sales, but I would feel a lot cheaper about myself.

With this guide, you are learning something and have been given facts, things you can check on for yourself and prove very easily. That's how science works my friend.

When you get your telescope, and set it up outside for the first night of viewing, keep in mind all those who have come before you. Astronomers call the first night out with a new scope 'first light'. Meaning its the first light that has passed through the telescope into their eyes.

When I get a first light situation with a telescope, I keep things very simple. Me, a stool for comfort, a warm jacket if its a cold night, some hot coffee, and the stars.

And always, looking over my shoulder with me I feel is the spirit of Carl Sagan, watching in silent appreciation for what I am seeing. To him and his memory I dedicate this book, and to the future Sagans out there, waiting for that first light and to have the fire of knowledge awakened within them.

Thank you for following along with me, and for giving me the chance to show you one of my favorite things.

--- Keith Nichols
    December 2012

www.ingramcontent.com/pod-product-compliance
Lightning Source LLC
Chambersburg PA
CBHW072046190526
45165CB00018B/1850